The Open University

Mathematics: A Second Level Course

Linear Mathematics   Unit 9

DIFFERENTIAL EQUATIONS II

Prepared by the Linear Mathematics Course Team

18-6-72

The Open University Press

The Open University Press Walton Hall Bletchley Bucks

First published 1972
Copyright © 1972 The Open University

Designed by the Media Development Group of the Open University.

Printed in Great Britain by
Martin Cadbury Printing Group

SBN 335 01098 9

This text forms part of a series of units that makes up the correspondence
element of an Open University Second Level Course. The complete list
of units in the course is given at the end of this text.

For general availability of supporting material referred to in this
text, please write to the Director of Marketing, The Open University,
Walton Hall, Bletchley, Buckinghamshire.

Further information on Open University courses may be obtained from
the Admissions Office, The Open University, P.O. Box 48, Bletchley,
Buckinghamshire.

1.1

Contents

Set Books

D. L. Kreider, R. G. Kuller, D. R. Ostberg and F. W. Perkins, *An Introduction to Linear Analysis* (Addison-Wesley, 1966).

E. D. Nering, *Linear Algebra and Matrix Theory* (John Wiley, 1970).

It is essential to have these books; the course is based on them and will not make sense without them.

Conventions

Before working through this correspondence text make sure you have read *A Guide to the Linear Mathematics Course*. Of the typographical conventions given in the Guide the following are the most important.

The set books are referred to as:

> **K** for *An Introduction to Linear Analysis*
> **N** for *Linear Algebra and Matrix Theory*

All starred items in the summaries are examinable.

References to the Open University Mathematics Foundation Course Units (The Open University Press, 1971) take the form *Unit M100 3, Operations and Morphisms*.

9.0 INTRODUCTION

In this unit we return to the study of linear differential equations, which we started in *Unit 4, Differential Equations I*. There we looked at linear differential equations from a general point of view, and saw how to reduce a very general type of first-order differential equation to the evaluation of integrals. For equations of higher order, there is no such general method for obtaining analytic solutions, but there are methods for dealing with special types of equation which often arise in applied mathematics. There are also general results yielding information which can be very useful, even though it may fall short of a complete solution.

The most fundamental of these results is the one we have seen already; that is, that the problem of finding the general solution of a linear differential equation of the form

$$Ly = h$$

where h is a known function and L is a linear differential operator, can be split into two parts. One part is to find the general solution y_h of the associated homogeneous equation

$$Ly = 0;$$

the second is to find a particular solution y_p of the original equation. Then it follows from the general theory of linear problems that $y_h + y_p$ is the general solution of the original equation.

In the present unit, we tackle the first of these parts: finding the general solution of a homogeneous differential equation of the form $Ly = 0$, where L is a linear differential operator of order at least 2. We shall deal first with a special, but very useful, type of homogeneous equation for which there is a recipe for writing down the general solution in all cases. These are the equations with constant coefficients. Afterwards we shall deal with homogeneous equations in general. Nonhomogeneous equations will be dealt with in *Unit 11, Differential Equations III: Nonhomogeneous Equations*.

The key idea is the fact that the solution set of $Ly = 0$ is a vector space whose dimension is equal to the order of L. This has the consequence that if we can find n linearly independent solutions of $Ly = 0$ (where n is the order of L) then these n solutions form a basis for the solution space. The entire solution space is therefore the set spanned by these n solutions, and the general solution is an arbitrary linear combination of them. We saw an example of this in the Foundation Course (*Unit M100 31, Differential Equations II*), where we studied the differential equation

$$X''(t) + X(t) = 0 \qquad (t \in [0, \infty))$$

which describes the free vibrations of a mass-spring system with suitable values for the mass and the stiffness of the spring.

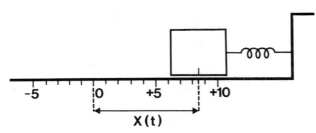

In the notation of this course (using y in place of X), the equation can be written

$$y'' + y = 0$$

or

$$(D^2 + 1)y = 0$$

where D is the differential operator. Since the differential operator $D^2 + 1$ is of second order, the solution space has two dimensions. Two linearly independent solutions of the equation are the sine and cosine functions, and since they are linearly independent, they form a basis for the solution space. The solution space is therefore

$$\langle \cos, \sin \rangle$$

and the general solution (an arbitrary element of this space) is the function

$$c_1 \cos + c_2 \sin$$

i.e.

$$t \longmapsto c_1 \cos t + c_2 \sin t \quad (t \in [0, \infty)),$$

where c_1 and c_2 are arbitrary numbers, the so-called arbitrary constants.

Whatever the differential equation, provided it is linear and homogeneous, the method of approach is the same: find n different solutions, check that they are linearly independent, and construct the general solution from them. If the equation has constant coefficients (like the equation $y'' + y = 0$ mentioned above) then it is always possible to find the necessary n solutions and hence solve the equation: this is discussed in Section 1 of this unit. If the equation does not have constant coefficients (as, for example, in the equation $y'' + xy = 0$), and the order is 2 or greater, then it is not, *in general*, possible to find explicit solutions; but it is still possible to find out a lot about the solutions. In Section 2 of the unit we shall see, in particular, how to prove the principle theoretical result for these equations, that the dimension of the solution space is equal to the order of the equation; we also obtain a method for checking that a given set of n solutions of the equation really are linearly independent and therefore can be used as a basis.

The first section of the unit is based on Sections 4-1, 4-2 and 4-3 of **K**; the second is based on Sections 3-4, 3-5 and 3-6. We reverse **K**'s order of presentation in order to put the easier material first, but if you prefer to treat general theorems before particular cases, there is no objection to your studying the second section of the unit before the first.

9.1 CONSTANT-COEFFICIENT EQUATIONS

9.1.1 Polynomials in D

As we saw in *Unit 4, Differential Equations I*, the general homogeneous linear differential equation has the form

$$a_n(x)y^{(n)}(x) + a_{n-1}(x)y^{(n-1)}(x) + \cdots$$
$$+ a_1(x)y'(x) + a_0(x)y(x) = 0 \qquad (y \in C^n(I))$$

where I is some interval of R and $a_n, a_{n-1}, \ldots, a_1, a_0$ are continuous real functions with domain I, and a_n is not the zero function. The functions $a_n, a_{n-1}, \ldots, a_0$ are called the coefficients, and if they are constant functions, then the equation is said to have constant coefficients. The significance of the fact that the coefficients are constants is indicated in the first reading passage.

READ page **K**126 *as far as* "... with relative ease".

Notes

(i)   *Equation (4-1)*   In this unit we shall be concerned only with the case $h = 0$.
(ii) *line 4*   Only a_n has to be nonzero (if it were 0, the order of the equation would be less than n).
(iii) *line 11*   "normal form" simply means that we divide the whole equation by a_n (which we know is nonzero) and then re-define $a_{n-1}, \ldots, a_0$ to stand for the quantities previously represented by $a_{n-1}/a_n, \ldots, a_0/a_n$.
(iv) *line −9*   "Algebraically such operators ..." The manipulation of these operators is considered in more detail in Example 4 on pages K51–52.

The relationship between constant-coefficient linear operators and polynomials is very important so we will go into it in a little more detail than **K**. Given such a linear operator

$$L = D^n + a_{n-1}D^{n-1} + \cdots + a_0$$

we associate with it a polynomial P such that

$$P(x) = x^n + a_{n-1}x^{n-1} + \cdots + a_0 \qquad (x \in R)$$

In this way we obtain a mapping M from the set of constant-coefficient linear differential operators (CCLO) with addition, scalar multiplication and the (composition) product to the set of polynomials with formal sums, scalar products and products, which is an isomorphism.

$$\text{CCLO} \underset{M^{-1}}{\overset{M}{\rightleftarrows}} \text{Polynomials}$$

For the product of differential operators in general we cannot "multiply out" as we do for polynomials, as can be seen in the example on page **K**88:

$$(xD + 2)(2xD + 1) = 2x^2D^2 + 7xD + 2$$

whereas straightforward formal polynomial multiplication would give

$$2x^2D^2 + 5xD + 2$$

If L_1, L_2 are constant-coefficient operators the situation is much nicer. If the corresponding polynomials are P_1, P_2, we can write

$$L_1 = M^{-1}(P_1), L_2 = M^{-1}(P_2)$$

and then $L_1 + L_2$ corresponds to the polynomial $P_1 + P_2$, i.e. $L_1 + L_2 = M^{-1}(P_1 + P_2)$;

the operator aL_1 corresponds to the polynomial aP_1, i.e. $aL_1 = M^{-1}(aP_1)$;

and the operator $L_1 \circ L_2 = L_1 L_2$ corresponds to $P_1 P_2$, i.e. $L_1 L_2 = M^{-1}(P_1 P_2)$.

Hence we have the following commutative diagrams (indicating the presence of a morphism)

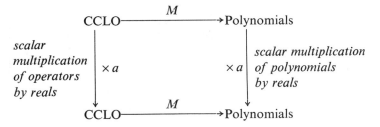

and the two diagrams

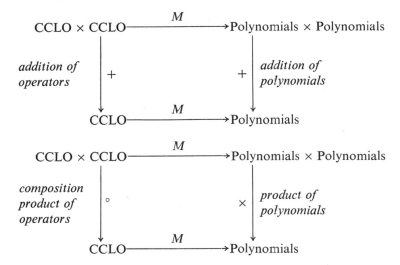

For example

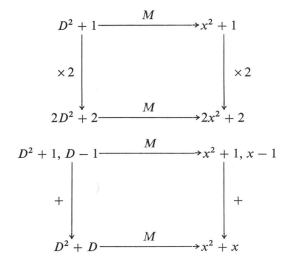

since

$$[(D^2 + 1) + (D - 1)]y = (D^2 + 1)y + (D - 1)y$$
$$= D^2y + y + Dy - y$$
$$= D^2y + Dy$$
$$= (D^2 + D)y$$

Exercise

If L_1 and L_2 are constant-coefficient operators, show that $L_1 \circ L_2 = L_2 \circ L_1$ by appealing to the mapping

$$M: \text{CCLO} \longrightarrow \text{Polynomials}$$

Solution

Since M is an isomorphism of (CCLO, $\circ$) to (Polynomials, $\times$) and multiplication of polynomials is commutative, it follows that $L_1 \circ L_2 = L_2 \circ L_1$.

Exercise

Draw the commutative diagram associated with the (composition) product of $(D^2 + 1)$ and $(D - 1)$.

Solution

$$
\begin{array}{ccc}
D^2 + 1,\, D - 1 & \xrightarrow{\quad M \quad} & x^2 + 1,\, x - 1 \\
\Big\downarrow {\scriptstyle \circ} & & \Big\downarrow {\scriptstyle \times} \\
D^3 - D^2 + D - 1 & \xrightarrow{\quad M \quad} & x^3 - x^2 + x - 1
\end{array}
$$

Since

$$
\begin{aligned}
[(D^2 + 1)(D - 1)]y &= (D^2 + 1)[(D - 1)y] \\
&= (D^2 + 1)(Dy - y) \\
&= D^2(Dy - y) + Dy - y \\
&= D^3 y - D^2 y + Dy - y \\
&= (D^3 - D^2 + D - 1)y
\end{aligned}
$$

For the reasons mentioned in the reading passage, it is important in this kind of work to know how to factorize a polynomial into linear and quadratic factors.

Example

Factorize the operators $D^2 - 8$, $D^3 - 8$.

As explained in the text following note (iv), the problem is the same as the problem of factorizing the polynomials $x^2 - 8$, $x^3 - 8$. Now any polynomial $P(x)$ of degree n can be factorized into the form

$$P(x) = a_n(x - \alpha_1)(x - \alpha_2) \cdots (x - \alpha_n) \tag{1}$$

where a_n is the coefficient of x^n and $\alpha_1, \alpha_2, \ldots, \alpha_n$ are real or complex numbers called the roots of $P(x) = 0$. They can be found by using the fact that every root is a solution of $P(x) = 0$.

For the polynomial $x^2 - 8$, we solve

$$x^2 - 8 = 0.$$

The roots are $\pm\sqrt{8}$, so the factorization of $x^2 - 8$ is

$$x^2 - 8 = (x - \sqrt{8})(x + \sqrt{8})$$

and the factorization of $D^2 - 8$ is $(D - \sqrt{8})(D + \sqrt{8})$.

For $x^3 - 8$, we must solve

$$x^3 - 8 = 0.$$

One root is obviously $x = \sqrt[3]{8} = 2$, and so there is a factor $x - 2$. To find the others we try

$$x^3 - 8 = (x - 2)(ax^2 + bx + c)$$

(there are two more factors; so their product will be a quadratic).

Multiplying out the right-hand side gives

$$x^3 - 8 = ax^3 + (b - 2a)x^2 + (c - 2b)x - 2c$$

We can regard the x^3, x^2, x and 1 as a basis for the vector space P_4, of polynomials of degree three or less. Then the two sides of the equation are apparently two different ways of representing the same polynomial in terms of this basis. But we know that any element of a vector space can be expressed uniquely in terms of a basis: so we can equate corresponding coordinates ("equate coefficients of like powers of x", as it is usually put in this context), i.e.

$$1 = a$$
$$0 = b - 2a$$
$$0 = c - 2b$$
$$-8 = -2c$$

Solving these in succession gives $a = 1$, $b = 2$, $c = 4$ and again $c = 4$, and putting these numbers into $ax^2 + bx + c$ in the previous factorization gives

$$x^3 - 8 = (x - 2)(x^2 + 2x + 4) \tag{2}$$

To complete the factorization, we would solve

$$x^2 + 2x + 4 = 0.$$

If we apply the formula for quadratics to this, however, we find that both the roots are complex; thus the linear factors of $x^2 + 2x + 4$ are also complex. Such factors do not interest us here, however; they would not lead to a suitable factorization of $D^3 - 8$, because our definition of a linear differential operator (see page **K**126, line 4) requires the coefficients to be real. Thus, the simplest factors we can get for $D^3 - 8$ are the ones corresponding to Equation (2), namely

$$D^3 - 8 = (D - 2)(D^2 + 2D + 4).$$

Exercises

1. Exercises 13(a), (b), (c), (d), page **K**90. In (d) one root of the relevant polynomial equation is 2. "Irreducible" means that any further factorization would involve complex coefficients.
2. Exercises 5(a), (b), (d), page **K**127. In each case, at least one of the roots of the corresponding polynomial equation is an integer. Use the fact that any integer root of

$$x^n + a_{n-1}x^{n-1} + \cdots + a_1 x + a_0 = 0$$

   must be a (positive or negative) factor of a_0.

Solutions

1. (a) $D^2 - 3D + 2 = (D - 1)(D - 2)$
    (b) $2D^2 + 5D + 2 = 2(D + \frac{1}{2})(D + 2)$ or $(2D + 1)(D + 2)$
    (c) $4D^2 + 4D + 1 = 4(D + \frac{1}{2})^2$ or $(2D + 1)^2$
    (d) $D^3 - 3D^2 + 4 = (D - 2)^2(D + 1)$

    One root of $x^3 - 3x^2 + 4 = 0$ is 2, so one factor is $x - 2$.

    Trying $x^3 - 3x^2 + 4 = (x - 2)(ax^2 + bx + c)$
    $$= ax^3 + (b - 2a)x^2 + (c - 2b)x - 2c$$

    and equating coefficients of like powers of x gives

    $$a = 1; b - 2a = -3; c - 2b = 0; -2c = 4.$$

    Thus

    $$a = 1, b = -1, c = -2, c = -2$$

    again.

Hence, the other factor is $(x^2 - x - 2)$, which factorizes to $(x - 2)(x + 1)$ and we get

$$x^3 - 3x^2 + 4 = (x - 2)(x - 2)(x + 1).$$

2. (a) $D^3 + 4D^2 + 5D + 2 = (D + 1)^2(D + 2)$

One root of $x^3 + 4x^2 + 5x + 2 = 0$ is -1, so $x + 1$ is a factor.

By the method used in Solution 1(d), the other factor is $x^2 + 3x + 2$, which factorizes to $(x + 1)(x + 2)$ and we get

$$x^3 + 4x^2 + 5x + 2 = (x + 1)(x + 1)(x + 2)$$

(b) $D^3 - D^2 + D - 1 = (D - 1)(D^2 + 1)$

The quadratic $x^2 + 1 = 0$ has no real roots, so there are no simpler real factors.

(d) $D^4 - 5D^2 + 4 = (D - 1)(D + 1)(D - 2)(D + 2)$.

One root of $x^4 - 5x^2 + 4 = 0$ is 1, so a factor is $x - 1$. This gives

$$x^4 - 5x^2 + 4 = (x - 1)(x^3 + x^2 - 4x - 4).$$

One root of $x^3 + x^2 - 4x - 4 = 0$ is -1, and so
$$x^3 + x^2 - 4x - 4 = (x + 1)(x^2 - 4)$$
$$= (x + 1)(x - 2)(x + 2).$$

9.1.2 Second-order Equations

The next reading passage explains how the factorization technique of the preceding section enables us to solve linear constant-coefficient equations.

READ the whole of Section 4-2, starting on page **K**127.

Notes

(i) *line -3, page* **K**127 A *lemma* is a preliminary minor theorem. "Null space" means "kernel".

(ii) *line 1, page* **K**128 In other words, we prove that $L_i y = 0$ implies $(L_1 \ldots L_n)y = 0$, for $i = 1, \ldots, n$. As **K** remarks the proof is trivial: almost all that is being used is the commutativity of operator multiplication to get the L_i to the end.

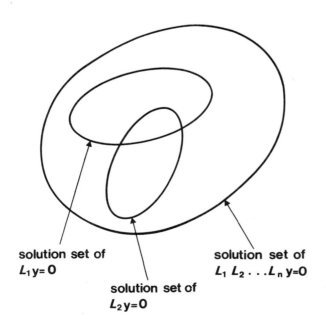

solution set of
$L_1 y = 0$

solution set of
$L_2 y = 0$

solution set of
$L_1 L_2 \ldots L_n y = 0$

(iii) *line 12, page* **K**128   The equation $(D - 2)y = 0$, i.e. $y' - 2y = 0$, could be solved by the method of separation of variables described on page **K**96; but it is enough just to check by differentiation that $y : x \longmapsto e^{2x}$ is a solution.

(iv) *line 13, page* **K**128   The Wronskian is a method of testing sets of functions for linear independence, which we shall study later in the unit. An alternative method that works here is this: if $c_1 e^{2x} + c_2 e^{-2x}$ is the zero function, then taking the particular values $x = 0$ and $x = 1$, we have $c_1 + c_2 = 0$ and $c_1 e^2 + c_2 e^{-2} = 0$, and it follows (since $e^2 \neq e^{-2}$) that $c_1 = c_2 = 0$.

(v) *lines 3 to 5, page* **K**129   Do not concern yourself here with the method by which the second solution $y(x) = xe^{\alpha x}$ is obtained; a method for obtaining this solution is given in *Unit 11, Differential Equations III*. The important point here is that if the auxiliary equation has equal roots, then a second solution of the differential equation is x times the first.

(vi) *line 9, page* **K**129   The reason why the method breaks down is that the $e^{\alpha_1 x}$ and $e^{\alpha_2 x}$ are now complex; consequently the functions

$$x \longmapsto e^{(a+bi)x}$$

and

$$x \longmapsto e^{(a-bi)x}$$

are not real functions, because their codomains are sets of complex numbers, rather than $\mathcal{R}$. Consequently, these functions do not belong to the domain of the linear operator $D^2 + a_1 D + a_2$ associated with the equation we are considering; for this domain is defined (on page **K**86) to be $\mathcal{C}^2(I)$, which is a set of real-valued functions, (i.e. functions with codomain $\mathcal{R}$). Of course we *could* have defined this domain to include complex-valued as well as real-valued functions, and had we done so functions such as $e^{(a+bi)x}$ would have been perfectly acceptable solutions of the equation. (There would still remain the problem, of course, of what we mean by differentiating such a function: but this can also be overcome reasonably easily.) It is often convenient, in fact, to extend the domain to include complex-valued functions and so take advantage of the greater simplicity of complex exponentials such as $e^{(a+bi)x}$ in comparison with expressions such as $e^{ax} \cos bx$. For the present, however, we stick to real functions in our rigorous arguments, and treat the argument based on $e^{(a+bi)x}$ as a piece of heuristic reasoning (i.e. plausible reasoning) which must be checked afterwards by rigorous methods. (There is a discussion, similar to that of *Case 3* in **K**, in *Unit M100 31, Differential Equations II*.)

(vii) *line 16, page* **K**129   Euler's formula was discussed in *Unit M100 29, Complex Numbers II*. Note that the symbol "x" in this formula corresponds to the "bx" of the rest of the calculation.

(viii) *line* -12, *page* **K**129   Here c_3 and c_4 are defined as $c_1 + c_2$ and $i(c_1 - c_2)$, respectively.

(ix) *line* -11, *page* **K**129   Mathematicians often use the word "formal" to mean that symbols are being manipulated without adequate regard for their meaning, i.e. for the rules governing these manipulations.

(x) *last line of table, page* **K**130   There is a small change of notation: the numbers denoted by c_3 and c_4 on the preceding page are denoted here by c_1 and c_2.

Examples

Find the general solution of the following differential equations

(a) $(D^2 - 2D - 3)y = 0$
(b) $(D^2 + 4D + 4)y = 0$
(c) $(D^2 + 6D + 10)y = 0$

Solutions

(a) We see that the operator factorizes

$$(D^2 - 2D - 3) = (D - 3)(D + 1)$$

and hence the auxiliary equation has distinct real roots 3 and -1. This gives us the general solution

$$y(x) = c_1 e^{3x} + c_2 e^{-x}.$$

(b) In this case the operator factorizes

$$(D^2 + 4D + 4) = (D + 2)^2$$

to give the repeated root -2 of the auxiliary equation. Hence, we have as a general solution

$$y(x) = (c_1 + c_2 x)e^{-2x}.$$

(c) In this case we cannot find real factors of the operator. We can use the formula to calculate the roots of the auxiliary polynomial equation:

$$\frac{-6 \pm \sqrt{36 - 40}}{2} = -3 \pm i$$

or we can observe that the operator is of the form $D^2 - 2aD + a^2 + b^2$ where $a = -3$ and $b = 1$. In either way we see that the roots of the auxiliary equation are $-3 \pm i$ and the general solution is

$$y(x) = c_1 e^{-3x} \cos x + c_2 e^{-3x} \sin x.$$

The fact that an operator of the form $D^2 - 2aD + a^2 + b^2$ corresponds to solutions of the form $e^{ax} \cos bx$, $e^{ax} \sin bx$ will be useful to remember for later exercises.

Exercises

1. Exercises 1, 4, 9 and 14, page **K**130. Check your solutions by substitution in the differential equation.
2. Exercises 18 and 20, page **K**130. (We met initial-value problems in *Unit 7, Recurrence Relations*.)

Solutions

1. **K**1.   Differential equation (DE) $(D^2 + D - 2)y = 0$

Auxiliary equation (AE) $m^2 + m - 2 = 0$

Solutions of AE $m = 1$ or -2

Solution of DE $y(x) = c_1 e^x + c_2 e^{-2x}$

K4.   Differential equation $(D^2 - 2D)y = 0$

Auxiliary equation $m^2 - 2m = 0$

Solutions of AE $m = 0$ or 2

Solution of DE $y(x) = c_1 + c_2 e^{2x}$ (since $e^0 = 1$)

K9.   Differential equation $(D^2 - 2D + 2)y = 0$

Auxiliary equation $m^2 - 2m + 2 = 0$

Solutions of AE $m = 1 + i, 1 - i$

(i.e. $a = 1, b = 1$)

Solution of DE $y(x) = e^x(c_1 \cos x + c_2 \sin x)$

K14.   Differential equation $(9D^2 + 6D + 1)y = 0$

Auxiliary equation $9m^2 + 6m + 1 = 0$

Solutions of AE $m = -\frac{1}{3}$ (twice)

Solution of DE $y(x) = (c_1 + c_2 x) \exp\left(-\frac{1}{3}x\right)$.

2. **K**18.   Differential equation $(4D^2 - 12D + 9)y = 0$

Solutions of AE $m = \frac{3}{2}$ (twice)

General solution of DE is $y = (c_1 + c_2 x) \exp\left(\frac{3}{2}x\right)$

Conditions $y(0) = c_1 = 1$

$y'(0) = c_2 + \frac{3}{2}c_1 = \frac{7}{2}$

whence $c_1 = 1, c_2 = 2$

Required solution of DE $y(x) = (1 + 2x) \exp\left(\frac{3}{2}x\right)$

K20.   Differential equation $(4D^2 - 4D + 5)y = 0$

Solutions of AE $m = \frac{1}{2} \pm i$

General solution of DE $y = e^{(1/2)x}(c_1 \cos x + c_2 \sin x)$

Conditions	$y(0) = c_1 = \frac{1}{2}$
	$y'(0) = \frac{1}{2}c_1 + c_2 = 1$
whence	$c_1 = \frac{1}{2},\ c_2 = \frac{3}{4}$
Required solution	
of DE	$y(x) = e^{(1/2)x}(\frac{1}{2} \cos x + \frac{3}{4} \sin x)$

9.1.3 Equations of Order Greater than 2

The method we have been using for second-order homogeneous equations generalizes in a straightforward way to equations of higher order. The method is described in the next reading passage.

READ the whole of Section 4-3, starting on page K132 omitting the proof of Theorem 4-1.

Notes

(i) *line 15, page K132* "these functions are linearly independent" Take this for granted at present; the authors return to the question of linear independence in Example 5 on page K135.

(ii) *line −8, page K135* As we have already remarked, the Wronskian is a method for testing linear independence of functions, which we shall look at later in the unit (sub-section 9.2.4).

(iii) *Equation (4-14), page K135* This is the usual strategy for proving linear independence: we show that the only linear combination of these functions that is equal to the zero function is the combination with $c_1 = c_2 = \cdots = c_6 = 0$.

(iv) *line −2, page K135* The symbol $\equiv$ (read "identically equals") is a way of indicating that this is a relation between functions, not numbers: the function $c_1(D + 5)^3(D^2 - 4D + 13)e^{2x}$ is equal to the zero function. K does not use this consistently: for instance, it could also be used in Equation (4-14).

(v) *line 13, page K136* To prove that c_5 and c_6 are 0, use the fact that e^{2x} is never 0, and consider the particular values $x = 0$ and $x = \frac{1}{6}\pi$ in the equation $c_5 \cos 3x + c_6 \sin 3x = 0$.

Exercises

1. Exercises 1,3,11 and 14, page K136.
2. Exercises 17(a) to (d), page K136. (*Hint* The necessary formulas are at the top of page K134.)
3. Find the solution of $y''' + 3y'' - y' - 3y = 0$ that satisfies the initial conditions $y(0) = 2,\ y'(0) = -2,\ y''(0) = 10$.

Solutions

1. **K1.** The DE is $(D^3 - 3D^2 - D - 3)y = 0$. Factorization by the method of sub-section 9.1.1 converts this to

$$(D + 3)(D + 1)(D - 1)y = 0$$

The general solution is therefore

$$y(x) = c_1 e^{-3x} + c_2 e^{-x} + c_3 e^{x}.$$

K3. The DE is $(4D^3 + 12D^2 + 9D)y = 0$, i.e.

$$4D(D - \tfrac{3}{2})^2 y = 0.$$

The general solution is therefore

$$y(x) = c_1 + (c_2 + c_3 x) \exp\left(-\tfrac{3}{2}x\right).$$

K11. The DE is $(D^4 + D^3 + D^2)y = 0$, i.e.

$$D^2(D^2 + D + 1)y = 0.$$

The roots of $m^2 + m + 1 = 0$ are $-\frac{1}{2} \pm \frac{1}{2}\sqrt{3}\,i$.

The general solution of the DE is

$$y(x) = c_1 + c_2 x$$
$$+ e^{-(1/2)x}(c_3 \cos \tfrac{1}{2}\sqrt{3}x + c_4 \sin \tfrac{1}{2}\sqrt{3}x)$$

K14. The DE is $(D^5 + 2D^3 + D)y = 0$, which factorizes to

$$D(D^4 + 2D^2 + 1)y = 0$$

or $D(D^2 + 1)^2 y = 0.$

The general solution is therefore

$$y(x) = c_1 + (c_2 + c_3 x)\cos x$$
$$+ (c_4 + c_5 x)\sin x.$$

2. K17. (a) Since $x^2 e^{x+1} = ex^2 e^x$, a suitable operator is $(D - 1)^3$ or $(D - 1)^3$ times any polynomial in D.

   (b) A suitable operator is $D^2 - 2aD + (a^2 + b^2)$ with $a = 2$, $b = 2$, i.e. $D^2 - 4D + 8$. (See example on page C12.)

   (c) $(D^2 + 1)^3$ annihilates both $2x^2 \sin x$ and $x \sin x$, and hence also their sum (or in fact any function $x \longmapsto p(x)\sin x$, where p is a polynomial of degree 2 or less).

   (d) D^2 annihilates $3 + 4x$ and $D + 2$ annihilates $-2e^{-2x}$. Their product, $D^2(D + 2)$ annihilates both and is a solution to the problem.

3. By Part 1 of Solution 1, the general solution of

$$y''' + 3y'' - y' - 3y = 0$$

is

$$y(x) = c_1 e^{-3x} + c_2 e^{-x} + c_3 e^x$$

The conditions are $y(0) = c_1 + c_2 + c_3 \quad = 2$
$$y'(0) = -3c_1 - c_2 + c_3 = -2$$
$$y''(0) = 9c_1 + c_2 + c_3 \quad = 10$$

The solution of these equations (use Gauss elimination) is $c_1 = 1$, $c_2 = 0$, $c_3 = 1$, and so the required solution of the differential equation is $y(x) = e^{-3x} + e^x$.

9.1.4 Damping and Stability of Vibrations

Linear constant-coefficient equations of the second order are very useful in the theory of mechanical and electrical vibrations. We saw in *Unit M100 31, Differential Equations II*, how the differential equation

$$mX''(t) + sX(t) = 0$$

with m and s positive numbers, provided a model of a mass-spring system. This equation can be made more realistic by including a term to describe the frictional forces. If we assume that the frictional force is proportional to the velocity $X'(t)$ and is in the opposite direction, then we can allow for this in the equation of motion by adding to the spring force $-sX(t)$, an additional frictional force $-kX'(t)$, where k is a positive number.

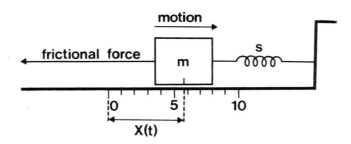

Thus the equation of motion becomes

$$mX''(t) = -sX(t) - kX'(t),$$

which is a second-order linear constant-coefficient equation. It can be written in the equivalent form

$$(mD^2 + kD + s)X = 0.$$

The auxiliary equation (using p for the polynomial variable instead of m, which is used for mass here) is

$$mp^2 + kp + s = 0$$

and this has roots which are

$$\frac{-k \pm \sqrt{k^2 - 4ms}}{2m}.$$

The methods given in sub-section 9.1.2, and summarized on page K130, show that the solutions of the differential equation can be of three types:

(i)   *small damping:* $0 < k^2 < 4ms$. The auxiliary equation has complex roots. In this case the general solution is

$$X(t) = \exp(-kt/2m)$$

$$\times \left(c_1 \cos \frac{\sqrt{4ms - k^2}}{2m} t + c_2 \sin \frac{\sqrt{4ms - k^2}}{2m} t \right)$$

Typical motions are indicated in the diagrams: the motion is oscillatory, as in the undamped case $k = 0$, but now the amplitude decreases with time. Mathematically, the oscillatory motion is provided by the cosine and sine terms, and the damping by the exponential term.

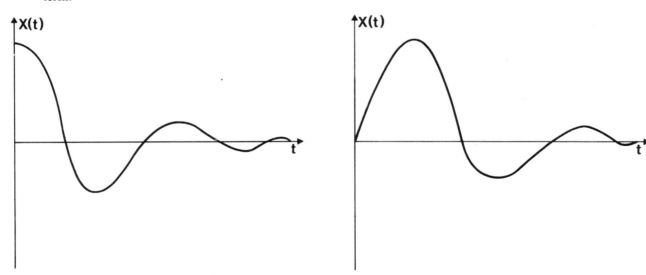

(ii)   *critical damping:* $k^2 = 4ms$. The auxiliary equation has equal roots, and the general solution is

$$X(t) = (c_1 + c_2 t) \exp\left(\frac{-k}{2m} t\right) = (c_1 + c_2 t) \exp\left(-\sqrt{\frac{s}{m}} t\right)$$

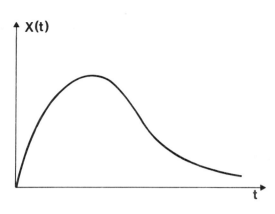

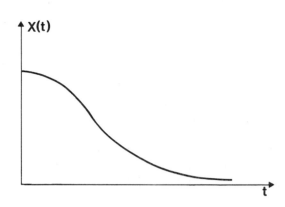

This time the motion is not oscillatory.

(iii)   *large damping:* $k^2 > 4ms$. The auxiliary equation has real roots, and the general solution is

$$X(t) = \exp\left(\frac{-kt}{2m}\right)$$

$$\times \left(c_1 \exp\frac{\sqrt{k^2 - 4ms}}{2m}\, t + c_2 \exp -\frac{\sqrt{k^2 - 4ms}}{2m}\, t\right)$$

Again the motion is not oscillatory.

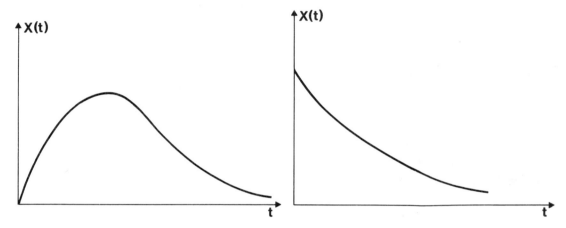

In each case the character of the motion depends crucially on the fact that k is *positive*. If k were negative, the factor $\exp(-kt/2m)$ would increase instead of decrease, and the motion would get more and more violent instead of dying away. This may seem unrealistic, but it can happen, as the following (optional) example shows.

Example (Optional)*

Consider the stability of the vibrations of an aircraft wing. In the analysis, we shall assume for simplicity that each section of the wing is elastically coupled, as if by a spring, to a fixed point (rather than to other parts of the wing). We also assume that the air moves past the wing in a horizontal direction (as in a wind tunnel) rather than the wing past the air. We denote the velocity of the air by V, and take it to be constant and positive.

There are two vertical forces on the wing. One is the elastic force, which depends on $X(t)$, the vertical displacement of the wing section from its equilibrium position. We denote this force by $F(X(t))$, with the convention that an upward force is positive. The second is the aerodynamic lift force, which depends on the angle of incidence α that the wing makes with the direction of the air flowing relative to it; we denote this lift force by $L(\alpha)$.

* This example ends on page 20.

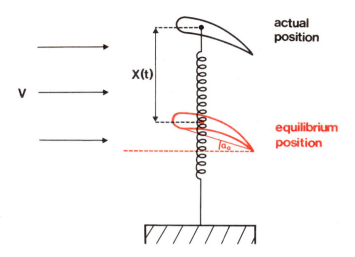

Now, if the wing is moving upwards with velocity $X'(t)$, the motion affects the angle of incidence; for during the time a particle of air travels a distance V to the right, the aerofoil travels a distance $X'(t)$ upwards and so the particle's displacement *relative to the aerofoil* is a combination of V to the right with $X'(t)$ downwards.

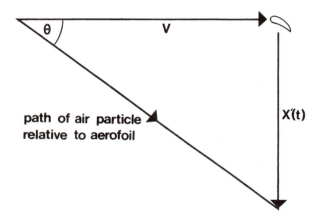

Thus, the velocity of the air stream relative to the aerofoil makes an angle θ with the horizontal, where

$$\tan \theta = \frac{X'(t)}{V}.$$

For small θ we can approximate* $\tan \theta$ by θ, so we have $\theta \simeq X'(t)/V$, and the angle of incidence α is given (see diagram) by

$$\alpha \simeq \alpha_0 - \frac{X'(t)}{V}$$

where α_0 is the angle between the aerofoil and the horizontal.

---

* This is the Taylor approximation of degree 1 at 0: see *Unit M100 14, Sequences and Limits II.*

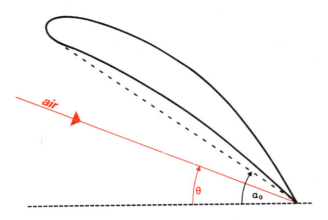

Combining the two vertical forces on the aerofoil, we obtain the equation of motion

$$mX''(t) = F(X(t)) + L\left(\alpha_0 - \frac{X'(t)}{V}\right),$$

where m is the mass of the aerofoil.

Let us make the assumption that F and L can be approximated by linear functions (i.e. polynomial functions of degree 1). We can use the Taylor approximation of degree 1 (see *Unit M100 14, Sequences and Limits II*), which for a general function f is

$$f(x) \simeq f(x_0) + (x - x_0)f'(x_0).$$

In our case x_0 is taken as the initial equilibrium position, when we have $X(0) = X'(0) = 0$.

We then have

$$F(X(t)) \simeq F(0) + X(t)F'(0)$$

and

$$L\left(\alpha_0 - \frac{X'(t)}{V}\right) \simeq L(\alpha_0) - L'(\alpha_0)\frac{X'(t)}{V}$$

Substituting these results in the differential equation, we get

$$mX''(t) = F(0) + F'(0)X(t) + L(\alpha_0) - L'(\alpha_0)\frac{X'(t)}{V}.$$

Since we are taking $X(t)$ to be the displacement *from equilibrium*, this equation must have the zero function as a solution; that is, we must have

$$0 = F(0) + L(\alpha_0)$$

The equation of motion thus simplifies to

$$mX''(t) = F'(0)X(t) - L'(\alpha_0)\frac{X'(t)}{V}$$

which we shall write as

$$mX''(t) + \frac{L'(\alpha_0)}{V}X'(t) + sX(t) = 0$$

where $s = -F'(0)$; this quantity is positive since an upward displacement of the aerofoil leads to a downward force on it from the spring.

The simplified equation of motion we have just obtained is of the linear type we have already investigated. As we have seen in this sub-section, the solutions contain a factor $\exp\left(-\dfrac{kt}{2m}\right)$ where k is the coefficient of $X'(t)$ in the differential equation. If this coefficient is positive, any oscillations will die away, but if it is negative, they will increase. Thus, the condition for the wing to be stable under this type of vibration is that k should be positive, i.e. (since V is positive)

$$L'(\alpha_0) > 0$$

The lift must increase, not decrease, with the angle of incidence. The function L normally looks like that in the diagram.

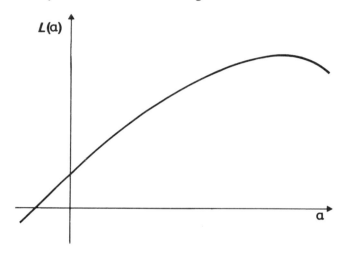

Thus stability can be achieved as long as the pilot keeps the angle of incidence within an acceptable range. On the other hand if $L'(\alpha_0)$ is negative (i.e. if the lift decreases with increasing angle of incidence) the vibrations will build up and may tear the wing apart.

Exercise

In designing systems that are capable of vibration it is often desirable to introduce friction into the system deliberately, to make the system return to equilibrium as quickly as possible after a disturbance.

If the system is modelled by the equation $mX'' + kX' + sX = 0$, the problem is to find the value of the friction constant k that gives the most rapidly decreasing solution for fixed m and s. The rate of decay is indicated by the term $e^{-\alpha t}$ in the formula for the solution, with the smallest value of α. (The larger α, the faster the decay; so we take the smallest value of α as the measure of decay, knowing that to be the slowest.)

Fill in the blanks in the following argument to show that decay is most rapid when there is critical damping, i.e. when $k^2 = 4ms$. For an exponential term $e^{-\alpha t}$, we shall call α its rate of decay.

For small damping, the rate of decay is (i) _____, but since in this case

$$0 < k^2 < \text{(ii)} \quad\rule{3cm}{0.4pt}$$

the rate of decay is less than (iii) _____.

For critical damping the rate of decay is

$$\text{(iv)} \rule{3cm}{0.4pt} = \text{(v)} \rule{3cm}{0.4pt}, \text{ as } k^2 = 4ms.$$

For large damping there are two exponential terms $e^{-\alpha_1 t}$, $e^{-\alpha_2 t}$ with $\alpha_1 \neq \alpha_2$, where $-\alpha_1$, $-\alpha_2$ are the roots of the auxiliary polynomial equation

(vi) _____. Hence

$$(p + \alpha_1)(p + \alpha_2) = \text{(vii)} \,\underline{\hspace{4cm}}$$

$$= \text{(viii)} \,\underline{\hspace{3cm}}$$

$$\text{and } \alpha_1 \alpha_2 = \text{(ix)} \,\underline{\hspace{2.5cm}}.$$

The rate of damping depends on the smaller of the terms α_1, α_2 and hence is less than (x) _____.

Of the three cases the highest rate of decay is for (xi) _____ damping.

Solution

(i) $\dfrac{k}{2m}$,   (ii) $4ms$,   (iii) $\sqrt{\dfrac{s}{m}}$,   (iv) $\dfrac{k}{2m}$,

(v) $\sqrt{\dfrac{s}{m}}$ (vi) $mp^2 + kp + s = 0$,

(vii) $p^2 + (\alpha_1 + \alpha_2)p + \alpha_1 \alpha_2$,

(viii) $p^2 + \dfrac{k}{m}p + \dfrac{s}{m}$

(ix) $\dfrac{s}{m}$

(x) $\sqrt{\dfrac{s}{m}}$

(xi) critical

9.1.5 Summary of Section 9.1

In this section we defined the terms

auxiliary equation	(page K128)	★ ★ ★
or characteristic equation	(page K128)	★ ★ ★
small damping	(page C16)	★
critical damping	(page C16)	★
large damping	(page C17)	★

We introduced the notation

M (page C7)

and saw that:
the function M from the set of all constant-coefficient linear differential operators to the set of polynomials is an isomorphism preserving addition, multiplication and scalar multiplication.

Techniques

1. Obtain a basis for the solution space of a constant-coefficient second-order equation.   ★ ★ ★

2. Specify the functions that factors of the form $(D - \alpha)^m$ and $(D^2 - 2aD + a^2 + b^2)^m$ contribute to the basis of the kernel of a constant-coefficient operator.   ★ ★ ★

9.2 EQUATIONS WITH VARIABLE COEFFICIENTS

9.2.0 Introduction

If the coefficients in the differential equation are not constant functions, then it is normally much harder to solve than the constant-coefficient type we considered in Section 1 of this unit. The general theory still applies, telling us that if we can find n linearly independent solutions of an nth order equation then we have a basis for the entire solution space; but these solutions are no longer just products of polynomial, trigonometric and exponential functions, and indeed they usually cannot be expressed in terms of elementary functions at all. In this section we obtain some general results that can be used to help overcome this kind of difficulty. These general results do not by themselves lead to a complete solution of the equation, but they do enable us to recognize a basis of the solution space when we see one, and sometimes to complete a basis when we have only some of its elements.

To recognize that a given set of functions, say $\{y_1, y_2, \ldots, y_n\}$, is a basis for the solution space of a given homogeneous differential equation $Ly = 0$, three steps must be taken. We must verify (i) that the given functions are in the solution space, (ii) that their number is equal to the dimension of the solution space and (iii) that they are linearly independent. These three conditions are sufficient to ensure that $\{y_1, \ldots, y_n\}$ is a basis, since we know that in a vector space of dimension n any n linearly independent vectors constitute a basis. The verification of the first of these three conditions is simply a matter of checking that the given functions satisfy the equation. In this section we will discuss how to verify the second and third.

We begin with the second condition. Here, it is a question of finding the dimension of the solution space. We have already used the fact that this dimension is equal to the order of the differential operator L; now we shall go into the reason why this is so.

9.2.1 First-order Equations

We begin with the simplest case: linear differential equations of order 1. We want to show that the solution space has dimension 1. We do this by mapping the solution space to another space which we know to have dimension 1, namely the space R. The solution space of such an equation is a family of functions, whose graphs can be sketched in the manner illustrated below (see *Unit M 100 24, Differential Equations I*). In the diagram the black lines are tangents to the solution curves for $y' = 2y$. Typical solution curves are shown in red.

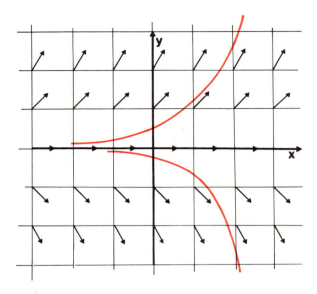

One way of mapping this family of curves to R is to map each curve to its intercept on the y-axis. More generally, we could map each curve to its intercept on any line parallel to the y-axis; say the line whose equation is $x = x_0$ with x_0 some real number.

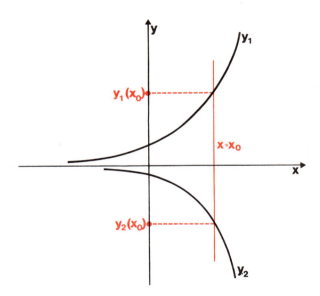

In symbols this mapping, which we denote by E, is

$$E : y \longmapsto y(x_0) \qquad (y \in S)$$

where S is the solution set of the first-order linear differential equation. It is not difficult to show that E is a linear transformation. (See Exercise 2 below.) E is called an *initial-condition mapping*.

To complete our argument we want to show that the linear transformation E *is* an isomorphism, i.e. that it is one-one and onto, or equivalently that it has an inverse. In other words we want to show that to every real number, say y_0, there corresponds a unique function y in the solution set such that $y(x_0) = y_0$. To do this we need a theorem, which is the subject of the next reading passage.

*READ from the beginning of Section 3-4, page **K**102 to the end of the statement of Theorem 3-1 page **K**103.*

Notes

(i) *line* -6, *page* **K102**   The equation under discussion is

$$a_1(x)y'(x) + a_0(x)y(x) = h(x) \qquad (x \in I)$$

which we studied in *Unit 4, Differential Equations I*. For the present you may take it that h is the zero function, but the results for other functions h will be useful later.

(ii) *Equation (3–30), page* **K102**   This was obtained in the preceding section of **K** (pages **K96–97**). You met it in *Unit 4*. The expression simplifies considerably since in our case $h = 0$.

(iii) *line 1, page* **K103**   "it is easy to see ...". In case you do not find it so easy, we give the formula for the solution curve (with $h = 0$) which passes through (x_0, y_0); it is

$$y = y_0 \exp\left(-\int_{x_0}^{x} \frac{a_0(x)}{a_1(x)}\, dx \right)$$

(Notice that when $x = x_0$, the integral is zero, and so we get $y = y_0$. Notice also the typical abuse of notation. The xs in $\frac{a_0(x)}{a_1(x)}\, dx$ are dummy variables for the purposes of integration and could be replaced by any other letter, whereas the x in $\int^{x}_{x_0}$ is the domain variable in I.) This shows that a solution *exists*; it remains to show that this solution is *unique*.

(iv) *line 9, page* **K103**   "governed by the equation" means here that the particle always moves in the direction whose slope is the value of $y'(x)$ given by the differential equation, i.e. the slope is (with $h = 0$)

$$-\frac{a_0(x)y(x)}{a_1(x)}$$

(v) *line* -10, *page* **K103**   "not difficult to show" In case you want to see a proof of this, here is one. We follow the suggestions in Exercise 14, page **K105**.

   (a)   Let y_1 and y_2 be two solutions of the equation: then subtracting using the formula (3-30) for the solution, we have

$$y_1(x) - y_2(x) = (c_2 - c_1) \exp\left(-\int \frac{a_0(x)}{a_1(x)}\, dx \right).$$

   The exponential part is never zero, so either $c_2 = c_1$, in which case $y_1 - y_2 = 0$ everywhere in I, or $c_2 \neq c_1$, in which case $y_1 - y_2 \neq 0$ everywhere in I.

   (b)   Suppose now that we have an initial-value problem, and suppose that we have solutions y_1 and y_2 (assumed distinct). Then

$$y_1(x_0) = y_0$$
$$y_2(x_0) = y_0$$

   and so at $x_0 \in I$, $y_1 - y_2$ is zero. Hence by part (a), $y_1 - y_2$ is zero everywhere on I, and the solutions are not distinct.

Example

Consider the equation $y' = y$.

Then the solution space is $S = \{x \longmapsto ce^x, c \in R\}$.

We define a mapping $E : S \longrightarrow R$, by

$$y \longmapsto y(1) \qquad (y \in S)$$

i.e. $(x \longmapsto ce^x) \longmapsto ce^1 = ce$.

This mapping is an isomorphism of the additive structures.

(i)   What are the images of $x \longmapsto e^x$, $x \longmapsto 2e^x$ under E?

(ii)   What are the inverse images of 1, 2?

(iii)   What is the inverse mapping $E^{-1} : R \longrightarrow S$?

(iv)   What is the solution of the equation $y' = y$, with initial condition $y(1) = 7$?

Solution

(i) By definition $E : (x \longmapsto e^x) \longmapsto e^1 = e$

$E : (x \longmapsto 2e^x) \longmapsto 2e^1 = 2e$

(ii) Since E sends $x \longmapsto e^x$ to e, it sends $x \longmapsto \dfrac{1}{e} e^x$ to 1.

But $\dfrac{1}{e} e^x = e^{(x-1)}$; hence the inverse image of 1 is $x \longmapsto e^{(x-1)}$ and similarly the inverse image of 2 is $x \longmapsto 2e^{(x-1)}$.

(iii) Extending (ii) we see that the inverse image of the number r is the function $x \longmapsto re^{(x-1)}$ and hence the inverse mapping E^{-1} is given by

$$E^{-1} : r \longmapsto (x \longmapsto re^{(x-1)}) \qquad (r \in R)$$

(iv) What does the mapping E^{-1} do? Well, it just gives us the solution of the equation with a given initial condition at 1. Hence if we want a solution with $y(1) = 7$, then all we need is $E^{-1}(7) = x \longmapsto 7e^{(x-1)}$.

Exercises

1. S denotes the solution set of the differential equation

$$y'(x) = 2xy(x) \qquad (x \in R)$$

and E is the function from S to R defined by

$$E : y \longmapsto y(2).$$

(i) Does the function $x \longmapsto \exp(x^2)$ belong to S?
(ii) Write down the image of this function under E.
(iii) Write down the inverse image of the number $\exp(4)$ under E.
(iv) If y_0 denotes an arbitrary real number, write down its inverse image under E.
(v) Write down the inverse of the function E.

2. Verify that E, as defined in the text on page C23, is a linear transformation.

Solutions

1. (i) Yes.
(ii) The image of y under E is defined to be $y(2)$; so if $y(x) = \exp(x^2)$ the image of this y is $\exp(4)$.
(iii) Since E maps the function $x \longmapsto \exp(x^2)$ to $\exp(4)$, the inverse image of $\exp(4)$ is this function.
(iv) The inverse image of the number y_0 is the solution of the differential equation such that $y(2) = y_0$. Since the general solution is $y(x) = c \exp(x^2)$, the required value of c is determined by $y_0 = c \exp(2^2)$, i.e. $c = y_0 e^{-4}$ and so the required solution is given by

$$y(x) = y_0 \exp(x^2 - 4) \qquad (x \in R).$$

The inverse image of y_0 is thus the function

$$x \longmapsto y_0 \exp(x^2 - 4) \qquad (x \in R).$$

(v) $E^{-1} : y_0 \longmapsto (x \longmapsto y_0 \exp(x^2 - 4)) \qquad (y_0 \in R).$

2. To show that E is a linear transformation we must show that its domain is a vector space and that

$$E(\alpha y + \beta z) = \alpha E(y) + \beta E(z).$$

The domain, S, is a vector space since it is the solution space of a homogeneous linear problem $Ly = 0$.

The transformation is linear since

$$E(\alpha y + \beta z) = (\alpha y + \beta z)(x_0) \quad \text{(definition of } E\text{)}$$
$$= \alpha y(x_0) + \beta z(x_0) \quad \text{(definitions of addition of functions and scalar multiplication)}$$
$$= \alpha E(y) + \beta E(z) \quad \text{(definition of } E\text{)}$$

9.2.2 Equations of Higher Order

To illustrate how the ideas explained in the last sub-section can be extended to higher-order equations, let us consider a second-order equation, say

$$y''(x) = xy(x) \qquad (x \in I) \tag{1}$$

We hope to show that its solution space S is two-dimensional, i.e. that it is isomorphic to R^2. One way of setting up the mapping from S to R^2 is to rewrite the equation as a pair of simultaneous first-order equations, in terms of y and a new function $z = y'$.

$$\begin{aligned} z'(x) &= xy(x) \\ y'(x) &= z(x) \end{aligned} \qquad (x \in I) \tag{2}$$

The solution set of this pair of equations is a set of pairs of functions of the form (y, z), and it is equivalent to the solution set S of Equation (1) in the sense that $y \in S$ if and only if the pair (y, y') is a solution of Equation (2).

The natural generalization of the initial-condition mapping E, which we looked at in the previous sub-section, to the system of simultaneous equations (2) is the mapping

$$(y, z) \longmapsto (y(x_0), z(x_0)) \qquad ((y, z) \in \text{solution set of Equation (2)})$$

where $x_0 \in I$. This mapping is equivalent to

$$(y, y') \longmapsto (y(x_0), y'(x_0)) \qquad (y \in S).$$

This idea leads to a natural way of mapping the solution space S of our original Equation (1) to R^2; as before we shall denote this initial-condition mapping by E:

$$E : y \longmapsto (y(x_0), y'(x_0)) \qquad (y \in S)$$

Once again, it is not hard to show that E is a linear transformation. To complete our proof that S is two-dimensional, we want to show that E is an isomorphism, i.e. that it has an inverse. This is done by proving that to every ordered pair of real numbers (y_0, y_0') there corresponds a unique function y in the solution set such that $y(x_0) = y_0$ and $y'(x_0) = y_0'$. The theorem which proves this is stated (but not proved) in the next reading passage.

READ from line -6, page K103 to line 13, page K104.

Notes

(i) *line -1, page K103* The proof referred to will be treated in *Unit 33, Existence and Uniqueness Theorem for Differential Equations.*

(ii) *Equation (3-32), page K104* The notation is a little different from ours: the book uses y_1 for our y_0', y_2 for our y_0'', etc.

The main result of the last two sub-sections can be summed up in the following theorem:

Theorem

If L is a normal linear differential operator of order n, with domain $C^n(I)$, S is the solution set of $Ly = 0$, and x_0 is any point in I, then the mapping E from S to R^n defined by

$$E : y \longmapsto (y(x_0), y'(x_0), \ldots, y^{(n-1)}(x_0))$$

is an isomorphism.

Example

Consider the differential equation

$$y''(x) + y'(x) = 0$$

and initial point $x_0 = 0$. The equation has solution space S with basis $\{x \longmapsto 1, x \longmapsto e^{-x}\}$. Let $E : S \longrightarrow R^2$ be the initial-condition mapping

$$E : y \longmapsto (y(x_0), y'(x_0)) \qquad (y \in S).$$

(i) What is

    (a) $E(x \longmapsto 1)$,

    (b) $E(x \longmapsto e^{-x})$,

    (c) $E^{-1}(1, 0)$,

    (d) $E^{-1}(0, 1)$?

(ii) What is $E^{-1}(c_1, c_2)$?

Solution

    (i)   (a). If $y : x \longmapsto 1$, then $y' : x \longmapsto 0$

$$E(y) = (y(x_0), y'(x_0)) = (y(0), y'(0)) = (1, 0)$$

        (b) If $y : x \longmapsto e^{-x}$, then $y' : x \longmapsto -e^{-x}$

$$E(y) = (y(0), y'(0)) = (1, -1).$$

        (c) From (a) we have $E^{-1}(1, 0) = x \longmapsto 1$

        (d) We can use the results in (a) and (b). The pairs $(1,0)$ and $(1, -1)$ are a basis for R^2 and

$$(0, 1) = (1, 0) - (1, -1).$$

Using the fact that E^{-1}, an isomorphism, is a linear transformation (see **Theorem 1.1,** *page* N28) we have

$$\begin{aligned}
E^{-1}(0, 1) &= E^{-1}(1, 0) - E^{-1}(1, -1) \\
&= (x \longmapsto 1) - (x \longmapsto e^{-x}) \\
&= x \longmapsto 1 - e^{-x}
\end{aligned}$$

    (ii) We can use the same technique as in (i) (d).

$$(c_1, c_2) = (c_1 + c_2)(1, 0) - c_2(1, -1)$$

so that $E^{-1}(c_1, c_2) = x \longmapsto c_1 + c_2 - c_2 e^{-x}$.

The function $E^{-1}(c_1, c_2)$ is just the solution of the original equation $y''(x) + y'(x) = 0$ which satisfies the initial conditions $y(0) = c_1$, $y'(0) = c_2$.

Exercises

1. For the differential equation

$$y''(x) + y(x) = 0 \qquad (x \in R)$$

with $x_0 = 0$, write down $E(x \longmapsto \cos x)$, $E(x \longmapsto \sin x)$, $E^{-1}(1, 0)$, $E^{-1}(0, 1)$ and $E^{-1}(c_1, c_2)$. Note that the solution space has basis $\{\cos, \sin\}$.

2. For the differential equation

$$y''(x) - y(x) = 0$$

with $x_0 = 0$, write down $E(x \longmapsto e^x)$, $E(x \longmapsto e^{-x})$, $E^{-1}(1, 0)$, $E^{-1}(0, 1)$, $E^{-1}(c_1, c_2)$. Note that the solution space has basis $\{x \longmapsto e^x, x \longmapsto e^{-x}\}$.

3. Exercise 2, page **K**105.

4. Exercise 15, page **K**105.

Solutions

1. $E(x \longmapsto \cos x) = (\cos 0, \cos' 0)$
    $\qquad\qquad\qquad = (1, 0)$, since $\cos' = -\sin$
    $E(x \longmapsto \sin x) = (\sin 0, \sin' 0)$
    $\qquad\qquad\qquad = (0, 1)$ since $\sin' = \cos$
    $\qquad E^{-1}(1, 0) = x \longmapsto \cos x$
    $\qquad E^{-1}(0, 1) = x \longmapsto \sin x$
    $\qquad E^{-1}(c_1, c_2) = x \longmapsto c_1 \cos x + c_2 \sin x$

    We can represent this solution by a diagram as in the T.V. programme

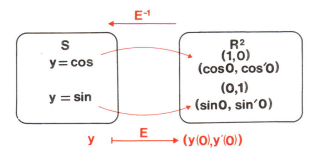

2. $\qquad E(x \longmapsto e^x) = (\exp 0, \exp' 0)$
    $\qquad\qquad\qquad = (1, 1)$
    $\qquad E(x \longmapsto e^{-x}) = (1, -1)$
    $\qquad E^{-1}(1, 0) = E^{-1}(\tfrac{1}{2}(1, 1) + \tfrac{1}{2}(1, -1))$
    $\qquad\qquad\qquad = x \longmapsto \tfrac{1}{2}(e^x + e^{-x})$
    $\qquad E^{-1}(0, 1) = E^{-1}(\tfrac{1}{2}(1, 1) - \tfrac{1}{2}(1, -1))$
    $\qquad\qquad\qquad = x \longmapsto \tfrac{1}{2}(e^x - e^{-x})$
    $\qquad E^{-1}(c_1, c_2) = x \longmapsto \tfrac{1}{2}c_1(e^x + e^{-x}) + \tfrac{1}{2}c_2(e^x - e^{-x})$

3. If we want the equation to be normal, we must choose an interval in which $\sin x$ does not vanish and which includes the initial value $\dfrac{3\pi}{4}$. Such an interval is $(0, \pi) = \{x: 0 < x < \pi\}$.

    In this interval we can apply standard techniques (see *Unit 4, Differential Equations I*) to obtain the general solution in the form

    $$y(x) = c/\sin x.$$

    Using the initial condition, we get

    $$y\left(\frac{3\pi}{4}\right) = 2 = c\sqrt{2}$$

    i.e.

    $$c = \sqrt{2}$$

    Thus, the solution to the initial value problem is

    $$y(x) = \sqrt{2}/\sin x \ (x \in (0, \pi)).$$

4. A solution to the exercise is given on page **K**733. Alternatively we could use the equation in Exercise 3 above. It is not normal

on the interval $[0, \pi) = \{x: 0 \leqslant x < \pi\}$, and we cannot find a solution for which $y(0) \neq 0$, because if we put $x = 0$ in the equation, we get

$$(\sin 0)y'(0) + (\cos 0)y(0) = 0$$

i.e. $$y(0) = 0.$$

So $y(0)$ cannot be specified arbitrarily.

9.2.3 Dimension of the Solution Space

The theorem stated at the end of the last sub-section gives the dimension of the solution space of a normal nth-order linear differential equation; for this solution space is isomorphic to R^n, which has dimension n, and isomorphic spaces have equal dimensions. This result is consolidated in the next reading passage.

READ Section 3-5 of **K**, *omitting the proof of* **Theorem 3-3** and the following paragraph.

Notes

(i) *line −7, page* **K106**   The $\mathcal{C}(I)$ in the statement of the theorem would have been better as $\mathcal{C}^n(I)$ see page **K92**. It is not wrong, however, because $\mathcal{C}^n(I)$ is a subspace of $\mathcal{C}(I)$. Over the next few pages the authors skip about a bit: it is always best to use the most appropriate subspace; "most appropriate" is determined by the highest derivative required in the context.

(ii) *line −6, page* **K106**   You need not read this proof, since the theorem follows immediately from the existence of the isomorphism E, as noted above.

(iii) *Example 1, page* **K108**   In this example, the standard basis in $\mathcal{R}^2$ is mapped into S and yields a basis in S. The functions forming this basis, denoted by cosh and sinh, are called the *hyperbolic cosine* and the *hyperbolic sine*.* You have already met them in Exercise 2 of the preceding sub-section, but they were not introduced properly. They have properties analogous to those of the cosine and sine functions; this analogy arises from the similarity between the formulas defining cosh and sinh and formulas for cos and sin which can be derived from Euler's formula (see *Unit M100 29, Complex Numbers II*).

$$\cosh x = \frac{1}{2}(e^x + e^{-x}), \quad \cos x = \frac{1}{2}(e^{ix} + e^{-x})$$

$$\sinh x = \frac{1}{2}(e^x - e^{-x}), \quad \sin x = \frac{1}{2i}(e^{ix} - e^{-ix}).$$

As an example of this analogy, we saw in Exercises 1 and 2 of the preceding sub-section that for the equation $y'' + y = 0$, we have $E^{-1}(1, 0) = \cos$ and $E^{-1}(0, 1) = \sin$; but, for $y'' - y = 0$, we have $E^{-1}(1, 0) = \cosh$ and $E^{-1}(0, 1) = \sinh$.

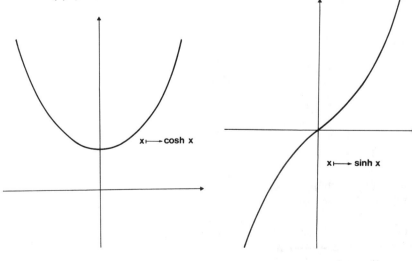

$x \longmapsto \cosh x$

$x \longmapsto \sinh x$

* The functions cosh and sinh are usually read as "cosh" and "shine", respectively.

(iv) *Example 2, page* **K109** This time we use $\{(1, 1), (1, -1)\}$ as our basis in $\mathscr{R}^2$, and the corresponding basis in S is $\{e^x, e^{-x}\}$.

(v) *Corollary 3-1, page* **K109** In terms of the isomorphism E, this corollary may be stated as follows: if B is a basis for $\mathscr{R}^n$, then $E^{-1}(B)$ is a basis for S.

Example

(i) Show that the functions $x \longmapsto \cosh x$, $x \longmapsto \sinh x$ and $x \longmapsto \cosh(a + x)$, where a is a constant, are all solutions of

$$(D^2 - 1)y = 0.$$

(ii) By considering $E: S \longmapsto R^2$ for initial point $x_0 = 0$, find a linear relation between them.

Solution

(i) From the formulas

$$\cosh x = \tfrac{1}{2}(e^x + e^{-x})$$
$$\sinh x = \tfrac{1}{2}(e^x - e^{-x}),$$

we find that $\cosh'(x) = \sinh x$ and $\sinh'(x) = \cosh x$. Hence

$$\cosh''(x) = \sinh'(x) = \cosh x$$
$$\cosh''(x) - \cosh x = 0$$
$$(D^2 - 1)\cosh = 0.$$

Similarly, $(D^2 - 1)\sinh = 0$

Since a is a constant, the function $x \longmapsto \cosh(a + x)$ also satisfies the differential equation.

$$E: S \longrightarrow R^2$$

is specified by $E: y \longmapsto (y(0), y'(0))$

Hence

$$E(x \longmapsto \cosh x) = (1, 0)$$
$$E(x \longmapsto \sinh x) = (0, 1)$$
$$E(x \longmapsto \cosh(a + x)) = (\cosh a, \sinh a)$$

(ii) Since E is an isomorphism, any linear relationship between

$$E(x \longmapsto \cosh x),$$
$$E(x \longmapsto \sinh x),$$
$$E(x \longmapsto \cosh(a + x))$$

corresponds to one between the functions $\cosh x$, $\sinh x$, $\cosh(a + x)$.

Now,

$$E(x \longmapsto \cosh(a + x))$$
$$= \cosh a \, E(x \longmapsto \cosh x)$$
$$+ \sinh a \, E(x \longmapsto \sinh x).$$

Hence, $\cosh(a + x) = \cosh a \cosh x + \sinh a \sinh x$.

Exercises

1. (i) Show that $x \longmapsto \sin(a + x)$, $x \longmapsto \sin x$, $x \longmapsto \cos x$ satisfy $(D^2 + 1)y = 0$.

   (ii) By considering the initial-condition mapping $E: S \longrightarrow R^2$ with $x_0 = 0$, find a linear relation between them.

2. (i) Show that the functions $x \longmapsto 1$, $x \longmapsto \cos 2x$, and $x \longmapsto \cos^2 x$ are solutions of $(D^3 + 4D)y = 0$.

   (ii) By considering the isomorphism $E: S \longrightarrow R^3$ with $x_0 = 0$, find any linear relations between them.

Solutions

1. (i) Differentiation shows that they satisfy the equation. Their images under E are

$$E(x \longmapsto \sin x) = (\sin 0, \sin' 0) = (0, 1)$$
$$E(x \longmapsto \cos x) = (\cos 0, \cos' 0) = (1, 0)$$
$$E(x \longmapsto \sin(a + x)) = (\sin a, \sin' a)$$
$$= (\sin a, \cos a)$$

(ii) In R^2 we have the linear relation

$$(\sin a, \cos a) = (\sin a)(1, 0) + (\cos a)(0, 1).$$

Applying E^{-1} to both sides and using the fact that it is in an isomorphism we find

$$E^{-1}(\sin a, \cos a) = \sin a\, E^{-1}(1, 0)$$
$$+ \cos a\, E^{-1}(0, 1)$$

i.e.

$$\sin(a + x) = \sin a \cos x + \cos a \sin x.$$

2. (i) Differentiation shows that they are solutions of the equation. Their images under E are

$$E(x \longmapsto 1) = (1, 0, 0)$$
$$E(x \longmapsto \cos 2x) = (\cos 0, -2 \sin 0, -4 \cos 0)$$
$$= (1, 0, -4)$$
$$E(x \longmapsto \cos^2 x) = (\cos^2 0, -2 \cos 0 \sin 0,$$
$$-2 \cos^2 0 + 2 \sin^2 0) = (1, 0, -2).$$

(ii) To find the linear relation between these vectors we can use inspection or the method of *Unit 3, Hermite Normal Form*, (pages N72 and N73). Using the latter method we consider the matrix whose columns are the coordinates of the 3 vectors,

$$\begin{bmatrix} 1 & 1 & 1 \\ 0 & 0 & 0 \\ 0 & -4 & -2 \end{bmatrix},$$

and by row operations reduce it to its Hermite normal form, which is

$$\begin{bmatrix} 1 & 0 & \frac{1}{2} \\ 0 & 1 & \frac{1}{2} \\ 0 & 0 & 0 \end{bmatrix}$$

The last column is half of the first plus half of the second and so the same relation holds for the columns of the first matrix

$$(1, 0, -2) = \tfrac{1}{2}(1, 0, 0) + \tfrac{1}{2}(1, 0, -4).$$

Applying E^{-1} to both sides of the equation we obtain

$$\cos^2 x = \tfrac{1}{2}(\cos 2x + 1)$$

a relation which is useful, for example, if you want to integrate $\cos^2 x$.

9.2.4 The Wronskian

For the general solution of a linear homogeneous equation it is not enough to have as many solutions as there are dimensions in the solution space; we must also make sure that these solutions are linearly independent, so that they form a basis. For example, in the last exercise of the preceding section, three solutions of the equation $(D^3 + 4D)y = 0$ are the functions $x \longmapsto 1$, $x \longmapsto \cos 2x$ and $x \longmapsto \cos^2 x$, but, since they are connected by the linear relation $1 + \cos 2x - 2\cos^2 x = 0$, they are not linearly independent, and so cannot be taken as a basis. In this subsection we generalize the idea of that exercise, using it to set up a general theoretical criterion for the linear independence of functions.

The criterion we shall set up is based on the determinant function, which you studied in *Unit 5, Determinants and Eigenvalues*. Remember that the determinant of any n vectors $\mathbf{a}_1, \ldots, \mathbf{a}_n$ in R^n is a scalar $D(\mathbf{a}_1, \ldots, \mathbf{a}_n)$, whose main property is that it vanishes if and only if the vectors are linearly dependent. We can also write the coordinates of $\mathbf{a}_1, \ldots, \mathbf{a}_n$ (with respect to the standard basis in R^n) as the columns of an $n \times n$ matrix, say A, and then $D(\mathbf{a}_1, \ldots, \mathbf{a}_n)$ is called the determinant of that matrix, written $\det A$. For example, in R^2, if $\mathbf{a}_1 = (p, q)$ and $\mathbf{a}_2 = (r, s)$ then we write

$$D(\mathbf{a}_1, \mathbf{a}_2) = \det \begin{bmatrix} p & r \\ q & s \end{bmatrix} = \begin{vmatrix} p & r \\ q & s \end{vmatrix}$$

$$= ps - rq.$$

Let us apply this idea to the three solutions of $(D^3 + 4D)y = 0$ just considered. We saw in the exercise that the images of the three solutions $x \longmapsto 1$, $x \longmapsto \cos 2x$, $x \longmapsto \cos^2 x$ under the isomorphism E are $(1, 0, 0)$, $(1, 0, -4)$ and $(1, 0, -2)$. The corresponding determinant is

$$\det \begin{bmatrix} 1 & 1 & 1 \\ 0 & 0 & 0 \\ 0 & -4 & -2 \end{bmatrix}, \text{ i.e. } \begin{vmatrix} 1 & 1 & 1 \\ 0 & 0 & 0 \\ 0 & -4 & -2 \end{vmatrix}$$

and it is equal to zero because one of its rows contains only zero entries. The vanishing of this determinant tells us that the three image vectors $(1, 0, 0)$, $(1, 0, -4)$ and $(1, 0, -2)$ are linearly dependent. Thus, since E is an isomorphism it follows that their inverse images, the functions $x \longmapsto 1$, $x \longmapsto \cos 2x$ and $x \longmapsto \cos^2 x$ are also linearly dependent, and do not constitute a basis for the solution space. Thus, if all we want to know is whether the functions constitute a basis, we can avoid the explicit calculation of the linear relations as we did in the exercise; the determinant tells us at once that *some* linear relation exists, and that is enough to rule out the given functions as a possible basis. On the other hand, if we had found the determinant to be different from zero, we would have known that its columns were linearly independent, and hence the original functions (which were known to span the space) would have been linearly independent and hence would have been suitable as a basis of the solution space.

Generalizing these considerations we have the following result:

Theorem 1

A set of n solutions $y_1, \ldots, y_n$ of an nth-order normal homogeneous linear differential equation is linearly independent if and only if the determinant

$$D(E(y_1), \ldots, E(y_n))$$

is not zero.

The proof is a consequence of the fact that E is an isomorphism: the functions $y_1, \ldots, y_n$ are linearly independent if and only if their images under E are linearly independent, and these are linearly independent if and only if the determinant of this set of image vectors is not zero.

There is another way of stating this theorem, which makes use of the fact that the mapping

$$E: y \longmapsto (y(x_0), y'(x_0), \ldots, y^{(n-1)}(x_0))$$

depends on the number x_0. This has the consequence that the determinant mentioned in the theorem depends on x_0; written out as the determinant of a matrix it looks like this:

$$D(E(y_1), \ldots, E(y_n)) = \det \begin{bmatrix} y_1(x_0) & \cdots & y_n(x_0) \\ y_1'(x_0) & \cdots & y_n'(x_0) \\ \vdots & & \vdots \\ y_1^{(n-1)}(x_0) & \cdots & y_n^{(n-1)}(x_0) \end{bmatrix}$$

This means that we can define a function (with domain I and codomain R):

$$W : x_0 \longmapsto \text{(the above determinant)} \qquad (x_0 \in I).$$

It is called the *Wronskian function* associated with the set of functions $y_1, \ldots, y_n$.

In terms of the Wronskian function we can re-state the above theorem in function language.

Theorem 2

A set of n solutions of an nth-order normal homogeneous linear differential equation on the interval I is linearly independent if and only if its Wronkskian function maps all the points in I to numbers other than zero; and it is linearly dependent if and only if its Wronskian function is the zero function.

Section **K3-6** gives some examples of the use of the Wronskian, beginning with cases where $y_1, \ldots, y_n$ are not necessarily solutions of a differential equation. In this case E is no longer an isomorphism, and so less can be proved; if the Wronskian function is ever non-zero, then the functions are linearly independent, but if it is the zero function, we can deduce nothing. We have not included any of this in the official reading passage, since we are mainly interested in the case where $y_1, \ldots, y_n$ do satisfy a differential equation.

READ (a) page **K**111 *down to the equation* "$W[x, 2x] \cdots = 0$"; *(b) the statements only of Theorems 3-5 and 3-6 on page* **K**113; *(c) the whole of page* **K**114.

Notes

(i) *Equation (3-41), page* **K**111 These vectors are just $E(y_1), \ldots, E(y_n)$.
(ii) *line 7, page* **K**114 "linearly independent in $\mathcal{C}(I)$" that is, the *functions* $\sin^3 x$ and $1/\sin^2 x$, both of which belong to $\mathcal{C}(I)$, are linearly independent.

Exercises

1. Examples 4 and 5, page **K**114 refer to an interval I in which $\tan x$ and $\cot x$ are both defined. Give an example of such an interval.

2. Use Theorem 2 to test the following sets of functions for linear independence, and write down a basis for the solution space if there is one among the functions given. In each case, take I to be R.

(i) e^x, e^{-x}	(consider $(D^2 - 1)y = 0$)
(ii) $\sin x, \cos x$	(consider $(D^2 + 1)y = 0$)
(iii) $1, x, x^2$	(consider $D^3 y = 0$)
(iv) $1 - x, x - x^2, x^2 - 1$	(consider $D^3 y = 0$)

Solutions

1. Any interval that includes no integer multiple of $\frac{\pi}{2}$ will do,

    for example, $\left(0, \frac{\pi}{2}\right)$ $\left(\text{but not } \left[0, \frac{\pi}{2}\right]\right)$.

    Since $\tan x = \dfrac{\sin x}{\cos x}$, it is defined unless $\cos x = 0$, i.e. unless

    x is an odd multiple of $\frac{\pi}{2}$. Since $\cot x = \dfrac{\cos x}{\sin x}$, it is defined

    unless x is an even multiple of $\frac{\pi}{2}$. Thus the interval must not

    include either even or odd multiples of $\frac{\pi}{2}$.

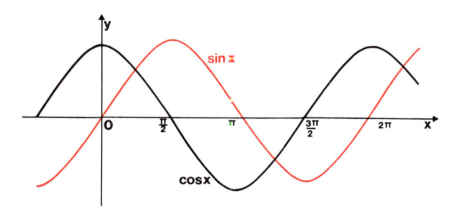

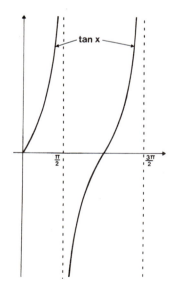

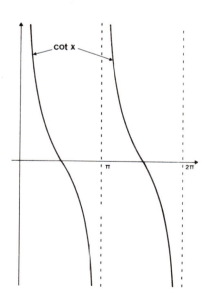

2.   (i)   Both functions satisfy the given differential equation, and

$$W = \begin{vmatrix} e^x & e^{-x} \\ e^x & -e^{-x} \end{vmatrix} = -2 \neq 0;$$

so they are linearly independent. Since $n = 2$ here, $\{e^x, e^{-x}\}$ is our basis.

(ii)   Both functions satisfy the differential equation, and

$$W = \begin{vmatrix} \sin x & \cos x \\ \cos x & -\sin x \end{vmatrix} = -\sin^2 x - \cos^2 x$$
$$= -1 \neq 0;$$

so they are linearly independent, and since $n = 2$, a basis is $\{\sin x, \cos x\}$.

(iii)   All three functions satisfy the equation, and

$$W = \begin{vmatrix} 1 & x & x^2 \\ 0 & 1 & 2x \\ 0 & 0 & 2 \end{vmatrix} = 2$$

since W is triangular.

Hence, the functions are linearly independent and since the equation is third-order, a basis is $\{1, x, x^2\}$. (In a similar way we can show that $\{1, x, x^2, \ldots, x^{n-1}\}$ is linearly independent, for any positive integer n.)

(iv)   All three functions satisfy the equation, and

$$W = \begin{vmatrix} 1-x & x-x^2 & x^2-1 \\ -1 & 1-2x & 2x \\ 0 & -2 & 2 \end{vmatrix}$$
$$= \begin{vmatrix} 1-x & x-1 & x^2-1 \\ -1 & 1 & 2x \\ 0 & 0 & 2 \end{vmatrix} \text{ (col. 2 + col. 3)}$$

$= 0$, since first 2 columns are proportional.

The given functions are linearly related, and do not form a basis.

9.2.5 Summary of Section 9.2

In this section we defined the terms

initial-condition mapping	(page C23)	★ ★ ★
initial-value problem	(page K103)	★ ★ ★
hyperbolic sine—sinh	(page C29)	★ ★ ★
hyperbolic cosine—cosh	(page C29)	★ ★ ★
uniqueness problem	(page K103)	★ ★ ★
Wronskian function	(page C33)	★ ★ ★

We introduced the notation

E	(page C23)	

Theorems

1. (**3.1**, page K103)
Every initial-value problem involving a normal first-order linear differential
equation has precisely one solution.   ★ ★ ★

2. (**3.2**, page K104) (The existence and uniqueness theorem for linear dif-   ★ ★ ★
ferential equations)

Let

$$a_n(x)\frac{d^n y}{dx^n} + \cdots + a_0(x)y = h(x) \tag{A}$$

be a normal nth order linear differential equation defined on an interval
I, and let x_0 be any point in I. Then if $y_0, \ldots, y_{n-1}$ are arbitrary real
numbers, there exists one and only one solution $y(x)$ of Equation (A) with
the property that

$$y(x_0) = y_0, \, y'(x_0) = y_1, \, \ldots, \, y^{(n-1)}(x_0) = y_{n-1}.$$

3. (page C27)   ★ ★ ★
If L is a normal linear differential operator of order n, with domain $C^n(I)$, S
is the solution set of $Ly = 0$, and x_0 is any point in I, then the mapping E
from S to R^n defined by

$$E: y \longmapsto (y(x_0), y'(x_0), \ldots, y^{(n-1)}(x_0))$$

is an isomorphism.

4. (**3-3**, page K106)
The solution space of a normal nth order linear homogeneous differential   ★ ★ ★
equation is n-dimensional.

5. (**Theorem 1**, page C32)
The solution $y_1, \ldots, y_n$ of an nth-order normal linear homogeneous
differential equation is linearly independent if and only if the determinant   ★ ★
$D(E(y_1), \ldots, E(y_n))$ is not zero.

6. (**Theorem 2**, page C33)
A set of n solutions of an nth-order homogeneous linear differential
equation on the interval I is linearly independent if and only if its   ★ ★
Wronskian function maps all the points in I to numbers other than zero;
and it is linearly dependent if and only if its Wronskian function is the
zero function.

Techniques

1. Use the initial condition mapping E and its inverse E^{-1} to solve initial   ★ ★ ★
value problems.
2. Use E to determine linear relations between functions in the solution   ★
space of a linear differential equation.
3. Use the Wronskian function to determine if a set of solutions of a   ★
linear differential equation forms a basis.

9.3 SUMMARY OF THE UNIT

The general aims of this unit are to show how to obtain the solution space for a constant-coefficient linear differential equation, given that the operator is factorized into first and second-order factors, and how to recognise a basis for the solution space of any homogeneous linear differential equation.

We began by showing that factorization of constant-coefficient differential operators is the same as that for polynomials and then showed how the solutions of a second order equation depended on whether the roots of the corresponding quadratic are real and distinct, real and the same or complex conjugates.

If the equation is $(D^2 + a_1 D + a_0)y = 0$ and

$$(D^2 + a_1 D + a_0) = (D - \alpha)(D - \beta)$$

then for

   (i) $\alpha \neq \beta$, both real, the general solution is $c_1 e^{\alpha x} + c_2 e^{\beta x}$
   (ii) $\alpha = \beta$, real, the general solution is $(c_1 + c_2 x)e^{\alpha x}$ and
   (iii) α, β complex, $\alpha = a + ib, b \neq 0$, the general solution is $e^{ax}(c_1 \cos bx + c_2 \sin bx)$.

Then we showed how these results could be extended to an operator of any degree, as long as we could factorize it. Typically the factor $(D - a)^r$ would contribute the functions $e^{ax}, \ldots, x^{r-1}e^{ax}$ to the solution space and the one of the form $(D^2 - 2aD + a^2 + b^2)^r$ would contribute $e^{ax} \cos bx$, $\ldots, x^{r-1}e^{ax} \cos bx, e^{ax} \sin bx, \ldots, x^{r-1}e^{ax} \sin bx$.

The first section ended with an example of modelling a physical situation, an aircraft wing, by differential equations, and discussed the relation between the coefficients of the equation and stability or damping.

In the second section we discussed general homogeneous linear differential equations and stated that for an nth order equation there exists a unique solution for each initial condition $y(x_0), \ldots, y^{(n-1)}(x_0)$. In this way we are able to set up the initial-condition mapping E from the solution space to R^n which is an isomorphism. This tells us that the solution space is n-dimensional and that if we have n functions in it which we wish to test for linear independence then we only have to test their images in R^n. Hence the functions $y_1, \ldots, y_n$ are independent if and only if $\det (E(y_1), \ldots, E(y_n))$ is not equal to zero. If they are linearly dependent then we can use Hermite normal form to discover any linear relations between them by finding the relations between their image vectors.

Definitions

	auxiliary equation	(page **K128**)	★ ★ ★
or	characteristic equation	(page **K128**)	★ ★ ★
	small damping	(page **C16**)	★
	critical damping	(page **C16**)	★
	large damping	(page **C17**)	★
	initial-condition mapping	(page **C23**)	★ ★ ★
	initial-value problem	(page **K103**)	★ ★ ★
	hyperbolic sine—sinh	(page **C29**)	★ ★ ★
	hyperbolic cosine—cosh	(page **C29**)	★ ★ ★
	uniqueness problem	(page **K103**)	★ ★ ★
	Wronskian function	(page **C33**)	★ ★ ★

Theorems

1. (**3.1**, page **K**103)
Every initial-value problem involving a normal first-order linear differential equation has precisely one solution.

2. (**3.2**, page **K**104)
(The existence and uniqueness theorem for linear differential equations)
Let

$$a_n(x)\frac{d^n y}{dx^n} + \cdots + a_0(x)y = h(x) \tag{A}$$

be a normal nth-order linear differential equation defined on an interval I, and let x_0 be any point in I. Then if $y_0, \ldots, y_{n-1}$ are arbitrary real numbers, there exists one and only one solution $y(x)$ of Equation (A) with the property that

$$y(x_0) = y_0, \, y'(x_0) = y_1, \ldots, y^{(n-1)}(x_0) = y_{n-1}.$$

3. (page **C**27)
If L is a normal linear differential operator of order n, with domain $C^n(I)$, S is the solution set of $Ly = 0$, and x_0 is any number in I, then the mapping E from S to R^n defined by

$$E: y \longmapsto (y(x_0), y'(x_0), \ldots, y^{(n-1)}(x_0))$$

is an isomorphism.

4. (**3-3**, page **K**106)
The solution space of a normal nth order linear homogeneous differential equation is n-dimensional.

5. (**Theorem 1**, page **C**32)
The solution $y_1, \ldots, y_n$ of an nth-order normal linear homogeneous differential equation is linearly independent if and only if the determinant $D(E(y_1), \ldots, E(y_n))$ is not zero.

6. (**Theorem 2**, page **C**33)
A set of n solutions of an nth-order homogeneous linear differential equation on the interval I is linearly dependent if and only if its Wronskian function maps all the points in I to numbers other than zero; and it is linearly dependent if and only if its Wronskian function is the zero function.

Techniques

1. Obtain a basis for the solution space of a constant-coefficient second order equation.
2. Specify the functions that factors of the form $(D - \alpha)^m$ and $(D^2 - 2aD + a^2 + b^2)^m$ contribute to the basis of the kernel of a constant-coefficient operator.
3. Use the initial-condition mapping E and its inverse E^{-1} to solve initial-value problems.
4. Use E to determine linear relations between functions in the solution space of a linear differential equation.
5. Use the Wronskian function to determine if a set of solutions of a linear differential equation forms a basis.

Notation

M	(page **C**7)
E	(page **C**23)

9.4 SELF-ASSESSMENT

Self-assessment Test

This Self-assessment Test is designed to help you test quickly your understanding of the unit. It can also be used, together with the summary of the unit for revision. The answers to these questions will be found on the next non-facing page. We suggest you complete the whole test before looking at the answers.

1 Which of the general solutions

   (a) $Ae^{-2x} + Be^{2x}$
   (b) $A + Be^{-2x}$
   (c) $Axe^{-2x} + Be^{-2x}$
   (d) $Ae^{-2x}\cos 2x + Be^{-2x}\sin 2x$

   corresponds to the following differential operators:

   (i) $D^2 + 4D + 8$
   (ii) $D^2 + 4D + 4$
   (iii) $D^2 + 2D$
   (iv) $D^2 - 4$

2 Write down a basis for the solution space for the following differential equations:

   (i) $(D - 1)(D - 4)(D + 5)y = 0$
   (ii) $(D + 1)^3 D^2 y = 0$
   (iii) $(D^2 + 1)(D - 2)y = 0$
   (iv) $(D^2 - 4D + 5)D^2 y = 0$
   (v) $(D^2 - 4D + 5)^3(D - 1)y = 0$
   (vi) $(D^2 + 1)(D^2 + 9)^2 y = 0$

3 Which of these functions occur as solutions to a homogeneous constant-coefficient second-order linear differential equation?

   (i) $x \longmapsto e^x$
   (ii) $x \longmapsto \cos x$
   (iii) $x \longmapsto xe^{2x}$
   (iv) $x \longmapsto x \sin x$
   (v) $x \longmapsto x^2 e^x$
   (vi) $x \longmapsto \cos 2x$
   (vii) $x \longmapsto \cos^2 x$

4 Find the simplest constant-coefficient linear differential operator which annihilates the following functions

   (i) $x \longmapsto x + \cos x - 3 \sin x$
   (ii) $x \longmapsto e^x + 3 + 4xe^x$
   (iii) $x \longmapsto (x + 1)(e^x - e^{-x})$

5 What is the dimension of the solution space of

   (i) $D^2 y = 0$
   (ii) $(D^2 + 3D)y = 0$
   (iii) $(D + 2)y = 0$?

6 Associated with the solution space $S = \{y : (D - 1)^2 y = 0\}$ and the initial point 0 we have the initial-condition mapping $E: S \longrightarrow R^2$. What is

   (i) $E(x \longmapsto e^x)$
   (ii) $E(x \longmapsto xe^x)$
   (iii) $E^{-1}(1, 0)$?

7 Given that $\cos 3x$, $\cos x$, $\cos^3 x$ are solutions of $(D^2 + 9)(D^2 + 1)y = 0$, find any linear relations between them.

Solutions to Self-assessment Test

1. We check which general solution corresponds to which differential operator either by seeing if the operator annihilates it or alternatively by factorizing the operator, constructing its general solution and then looking for it in the list.

(i) $D^2 + 4D + 8 = D^2 - 2aD + a^2 + b^2$ where $a = -2, b = 2$. Hence the basis of its solution space is $\{e^{-2x} \cos 2x, e^{-2x} \sin 2x\}$ and the general solution is (d).

(ii) $D^2 + 4D + 4 = (D + 2)^2$. The auxiliary polynomial has repeated root -2 and hence a basis for its solution space is $\{e^{-2x}, xe^{-2x}\}$. This corresponds to general solution (c).

(iii) $D^2 + 2D = D(D + 2)$. The auxiliary polynomial has roots $0, -2$. The solution space has basis $\{1, e^{-2x}\}$ and the general solution is (b).

(iv) $D^2 - 4 = (D - 2)(D + 2)$. The auxiliary polynomial has roots $2, -2$ and hence the solution space has basis $\{e^{2x}, e^{-2x}\}$ corresponding to general solution (a).

2. Since all the operators are already factorized as much as possible over the reals we can just write down the bases

(i) $\{e^x, e^{4x}, e^{-5x}\}$

(ii) $\{e^{-x}, xe^{-x}, x^2e^{-x}, 1, x\}$

(iii) $\{\cos x, \sin x, e^{2x}\}$

(iv) $\{e^{2x} \cos x, e^{2x} \sin x, 1, x\}$

(v) $\{e^{2x} \cos x, xe^{2x} \cos x, x^2e^{2x} \cos x, e^{2x} \sin x, xe^{2x} \sin x, x^2e^{2x} \sin x, e^x\}$

(vi) $\{\cos x, \sin x, \cos 3x, x \cos 3x, \sin 3x, x \sin 3x\}$.

3. (i) Yes: it is a solution of $(D - 1)Dy = 0$

(ii) Yes: it is a solution of $(D^2 + 1)y = 0$

(iii) Yes: it is a solution of $(D - 2)^2 y = 0$

(iv) No: as the lowest order homogeneous constant-coefficient equation it satisfies is $(D^2 + 1)^2 y = 0$, fourth order.

(v) No: as the lowest order homogeneous constant-coefficient equation it satisfies is $(D - 1)^3 y = 0$, third order.

(vi) Yes: it is a solution of $(D^2 + 4)y = 0$.

(vii) No: $\cos^2 x = \frac{1}{2}(1 + \cos 2x)$ and hence the lowest order homogeneous constant-coefficient equation it satisfies is $D(D^2 + 4)y = 0$, third order.

4. (i) x is annihilated by D^2, $\cos x$ and $\sin x$ by $(D^2 + 1)$. Hence $D^2(D^2 + 1)$ is the simplest annihilator of $x + \cos x - 3 \sin x$.

(ii) 3 is annihilated by D, e^x and xe^x by $(D - 1)^2$. Hence $D(D - 1)^2$ is the simplest annihilator of $e^x + 3 + 4x\, e^x$.

(iii) e^x, xe^x are annihilated by $(D - 1)^2$ and e^{-x}, xe^{-x} are annihilated by $(D + 1)^2$. Hence $(D - 1)^2(D + 1)^2 = (D^2 - 1)^2$ is the simplest annihilator of $(x + 1)(e^x - e^{-x})$.

5. The dimension of the solution space is just the order of the equation, so we have

(i) 2,   (ii) 2,   (iii) 1.

6. The initial-condition mapping for initial point 0 is just $E: S \longrightarrow R$

$$y \longmapsto (y(0), y'(0)) \qquad (y \in S).$$

Hence

(i) $E(x \longmapsto e^x) = (e^0, e^0) = (1, 1)$

(ii) $E(x \longmapsto xe^x) = (0e^0, (0 + 1)e^0) = (0, 1)$

and

(iii) $E^{-1}(1, 0) = E^{-1}((1, 1) - (0, 1))$

$\qquad\qquad = E^{-1}(1, 1) - E^{-1}(0, 1)$

$\qquad\qquad = e^x - xe^x = (1 - x)e^x$.

7. Given that $\cos 3x$, $\cos x$, $\cos^3 x$ are solutions of the fourth-order equation we consider the initial-condition mapping at the point 0:

$$E: S \longrightarrow R^4$$
$$y \longmapsto (y(0), y'(0), y''(0), y'''(0)) \qquad (y \in S),$$

where S is the solution space of the equation $(D^2 + 9)(D^2 + 1)y = 0$.
We find
$$E(x \longmapsto \cos 3x) = (1, 0, -9, 0)$$
$$E(x \longmapsto \cos x) = (1, 0, -1, 0)$$
$$E(x \longmapsto \cos^3 x) = (1, 0, -3, 0)$$

as $(\cos^3)'x = -3\cos^2 x \sin x \qquad\qquad (\cos^3)'(0) = 0$
$(\cos^3)''x = 6\cos x \sin^2 x - 3\cos^3 x \qquad (\cos^3)''(0) = -3$
$(\cos^3)'''x = -6\sin^3 x + 12\cos^2 x \sin x + 9\cos^2 x \sin x \; (\cos^3)'''(0) = 0$

Since E is an isomorphism any linear relation between $E(\cos 3x)$, $E(\cos x)$ and $E(\cos^3 x)$ corresponds to a linear relation between $\cos 3x$, $\cos x$ and $\cos^3 x$. It is easy to see that we only need consider relations between the vectors

$$\begin{bmatrix} 1 \\ -1 \end{bmatrix} \quad \begin{bmatrix} 1 \\ -3 \end{bmatrix} \quad \begin{bmatrix} 1 \\ -9 \end{bmatrix}$$

We reduce to Hermite normal form the array

$$\begin{bmatrix} 1 & 1 & 1 \\ -1 & -3 & -9 \end{bmatrix}$$

$$\begin{bmatrix} 1 & 1 & 1 \\ 0 & -2 & -8 \end{bmatrix}$$

$$\begin{bmatrix} 1 & 1 & 1 \\ 0 & 1 & 4 \end{bmatrix}$$

$$\begin{bmatrix} 1 & 0 & -3 \\ 0 & 1 & 4 \end{bmatrix}$$

Hence

$$\begin{bmatrix} 1 \\ 0 \\ -9 \\ 0 \end{bmatrix} = 4 \begin{bmatrix} 1 \\ 0 \\ -3 \\ 0 \end{bmatrix} - 3 \begin{bmatrix} 1 \\ 0 \\ -1 \\ 0 \end{bmatrix}$$

$$E\,(x \longmapsto \cos 3x) = 4E\,(x \longmapsto \cos^3 x) - 3E\,(x \longmapsto \cos x)$$

and since E is an isomorphism we have

$$\cos 3x = 4\cos^3 x - 3\cos x.$$

LINEAR MATHEMATICS

1 Vector Spaces
2 Linear Transformations
3 Hermite Normal Form
4 Differential Equations I
5 Determinants and Eigenvalues
6 NO TEXT
7 Introduction to Numerical Mathematics: Recurrence Relations
8 Numerical Solution of Simultaneous Algebraic Equations
9 Differential Equations II: Homogeneous Equations
10 Jordan Normal Form
11 Differential Equations III: Nonhomogeneous Equations
12 Linear Functionals and Duality
13 Systems of Differential Equations
14 Bilinear and Quadratic Forms
15 Affine Geometry and Convex Cones
16 Euclidean Spaces I: Inner Products
17 NO TEXT
18 Linear Programming
19 Least-squares Approximation
20 Euclidean Spaces II: Convergence and Bases
21 Numerical Solution of Differential Equations
22 Fourier Series
23 The Wave Equation
24 Orthogonal and Symmetric Transformations
25 Boundary-value Problems
26 NO TEXT
27 Chebyshev Approximation
28 Theory of Games
29 Laplace Transforms
30 Numerical Solution of Eigenvalue Problems
31 Fourier Transforms
32 The Heat Conduction Equation
33 Existence and Uniqueness Theorem for Differential Equations
34 NO TEXT